WATCHING THE WEATHER

When It's COLD

Written by
Noah Leatherland

All rights reserved.
Printed in India.

A catalogue record for this book is available from the British Library.

ISBN: 978-1-80505-598-3

Written by:
Noah Leatherland

Edited by:
Elise Carraway

Designed by:
Jasmine Pointer

©2024
BookLife Publishing Ltd.
King's Lynn, Norfolk
PE30 4LS, UK

FSC MIX Paper | Supporting responsible forestry
FSC® C195953

All facts, statistics, web addresses and URLs in this book were verified as valid and accurate at time of writing. No responsibility for any changes to external websites or references can be accepted by either the author or publisher.

AN INTRODUCTION TO BOOKLIFE RAPID READERS...

Packed full of gripping topics and twisted tales, BookLife Rapid Readers are perfect for older children looking to propel their reading up to top speed. With three levels based on our planet's fastest animals, children will be able to find the perfect point from which to accelerate their reading journey. From the spooky to the silly, these roaring reads will turn every child at every reading level into a prolific page-turner!

CHEETAH
The fastest animals on land, cheetahs will be taking their first strides as they race to top speed.

MARLIN
The fastest animals under water, marlins will be blasting through their journey.

FALCON
The fastest animals in the air, falcons will be flying at top speed as they tear through the skies.

Photo Credits
Images are courtesy of Shutterstock.com. With thanks to Getty Images, Thinkstock Photo and iStockphoto.
Cover – MeSamong, LAUDiseno, AlexanderTrou, koblizeek. Texture throughout – MeSamong. 4–5 – Pakhnyushchy, KAMONRAT, Studio_G. 6–7 – Kongpraphat, Andrei Stepanov, muratart. 8–9 – ch123, nadia_if, Anditya Creative. 10–11 – schankz, Frau aus UA, Aligator Pro, Receh Lancar Jaya. 12–13 – Aleksei Ignatov, Marinka Buronka. 14–15 – E Forafontova, BlueRingMedia. 16–17 – Just dance, EugeneEdge. 18–19 – movchanzemtsova, LightField Studios. 20–21 – CandyBox Images, Scherbinator, BRO.vector. 22–23 – Maria Sbytova, inspiring.team, YanLev Alexey.

CONTENTS

PAGE 4	What Is Weather?
PAGE 6	Temperature
PAGE 8	Snow
PAGE 10	Ice
PAGE 12	Hail
PAGE 14	Cold Winters
PAGE 16	Shivering
PAGE 18	Hypothermia
PAGE 20	Staying Safe
PAGE 22	Chilly Days
PAGE 24	Glossary and Index

WORDS THAT LOOK LIKE THIS ARE EXPLAINED IN THE GLOSSARY ON PAGE 24.

What Is WEATHER?

The weather is what it feels like outside. Lots of things can affect what the weather is like.

Sometimes, the weather can be cold and wet. Sometimes, it can be hot and dry. The weather is always changing.

TEMPERATURE

The temperature is the measure of how hot or cold something is. Temperature is a big part of weather.

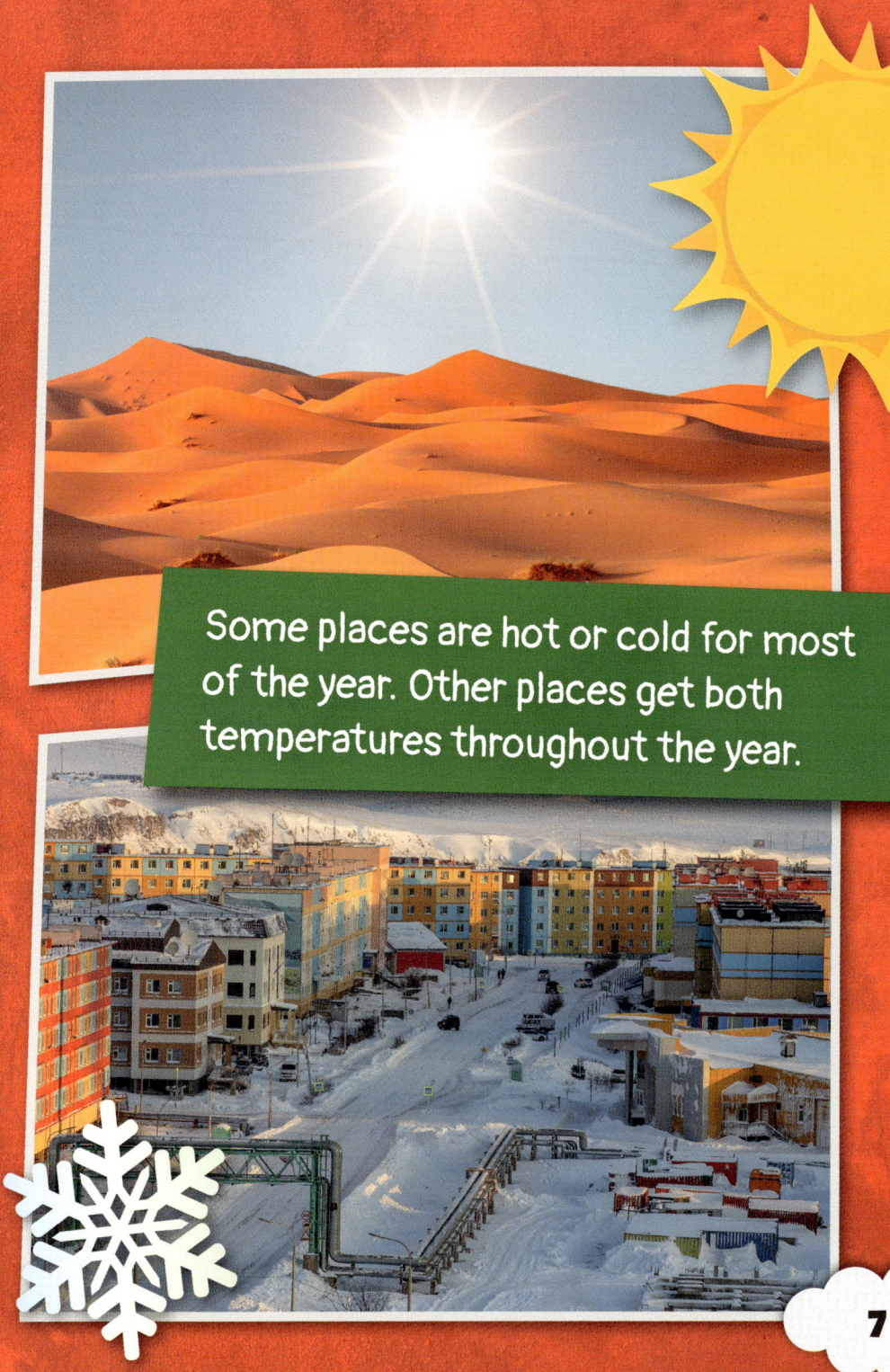

Some places are hot or cold for most of the year. Other places get both temperatures throughout the year.

SNOW

Snow is made up of tiny ice CRYSTALS. These crystals stick together in very cold clouds to make snowflakes.

Sometimes, snowflakes join together to make bigger snowflakes as they fall. Other times, snow can be quite loose and powdery.

ICE

Ice will form if the weather is cold enough. Water freezes if the temperature gets below zero **DEGREES CELSIUS**.

Ice can form on top of puddles and on the ground. Be careful, as it can be very slippery!

HAIL

Hail is made of water droplets that freeze in thunderclouds. As they get bigger and heavier, they start to fall.

Pieces of hail are called hailstones. Most hailstones are tiny, but they can be bigger than golf balls.

COLD WINTERS

Earth is slightly tilted. This means the top and bottom halves do not get the same amount of sunshine.

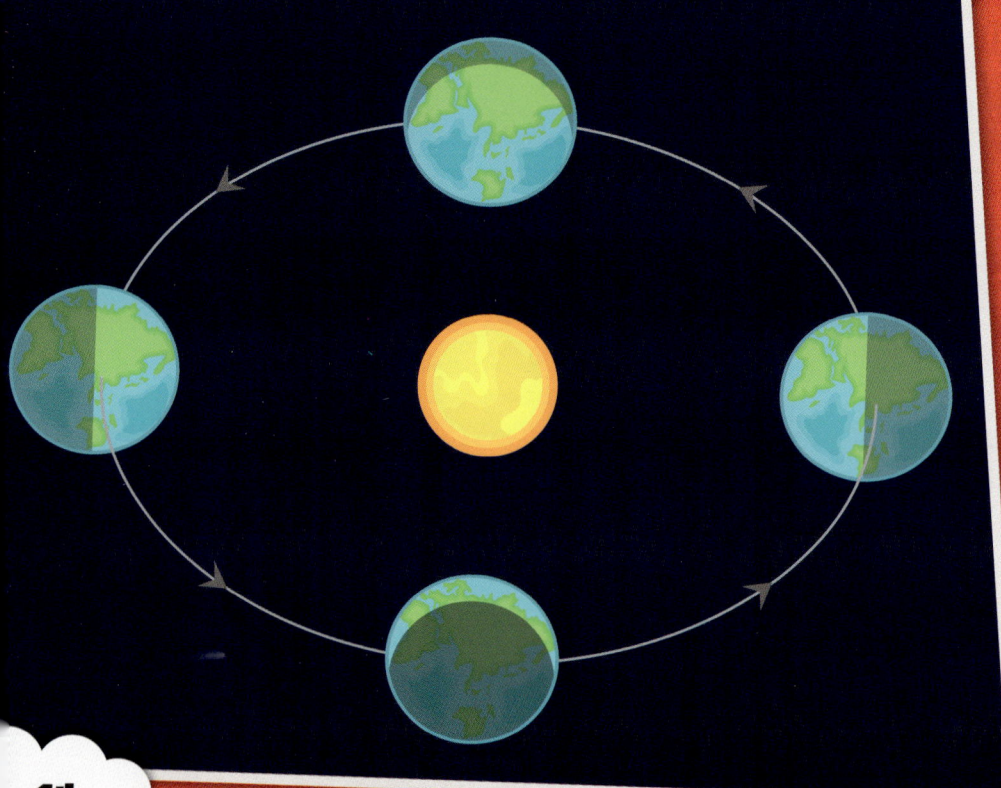

This tilt causes the seasons. It will be winter where the planet is tilted away from the Sun.

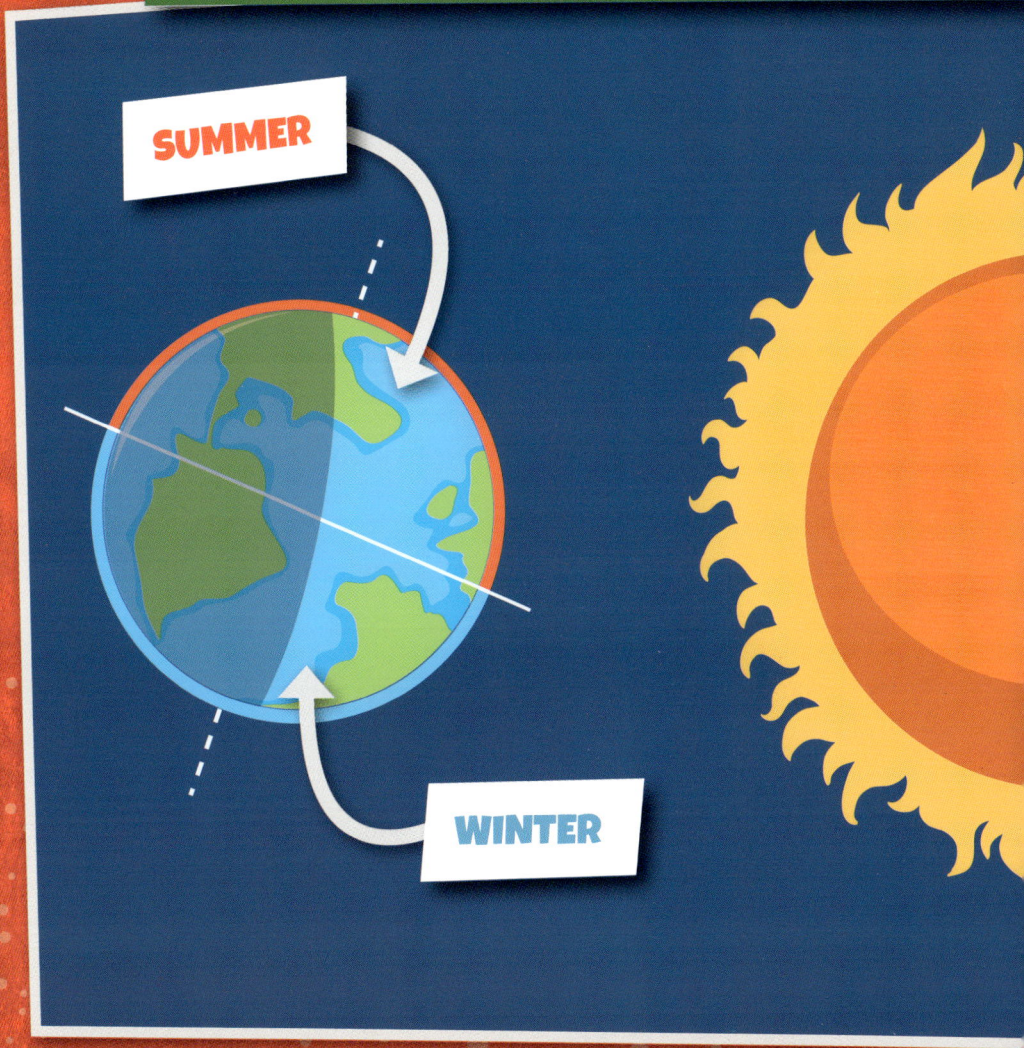

SHIVERING

Cold weather can make your body act differently. You might start to shiver if your body gets too cold.

Your **MUSCLES** tighten and relax very quickly when you shiver. Shivering is your body's way of warming itself up.

HYPOTHERMIA

Hypothermia is what happens when your body gets too cold. The human body cannot work properly when it has hypothermia.

Hypothermia can be easily avoided. Always wrap up warm before going out in the cold.

STAYING SAFE

It is important to stay safe in cold weather. Wrapping up warm is just one part of taking care of yourself.

Be extra careful around roads. Snow can make it hard to see. Ice can make roads slippery to cross.

CHILLY DAYS

Do not let the chilly weather put you off going outside. There is still lots you can do in the cold.

Grab your hat, scarf and gloves. Throw on your winter jacket and go and have some fun!

GLOSSARY

CRYSTALS — solid objects that are shaped into regular patterns

DEGREES CELSIUS — a scale of temperature where water freezes at zero degrees and boils at one hundred degrees

MUSCLES — the parts of the body that move the body around

INDEX

CLOUDS 8, 12
CRYSTALS 8
HAIL 12–13
ICE 8, 10–11, 21
ROADS 21

SEASONS 15
SNOW 8–9, 21
TEMPERATURES 6–7, 10
WATER 10, 12